Green Philanthropy For Families

160 Simple Earth Honoring Gifts, Actions, Activities and Projects

Helen, Justin, and Alexis Deffenbacher

Green Philanthropy For Families.Org Publications

ISBN: 978-0-615-33115-7
Library of Congress Control Number: 2009914189
Library of Congress subject heading: Environment

Cover: "First Garden," oil painting
by Debra Groesser: **www.debragroesser.com**

Published by:

Green Philanthropy For Families.Org Publications

www.greenphilanthropyforfamilies.org

DEDICATION

This book is dedicated to the families, individuals, groups and organizations all around the world that have joined the movement to build a more equitable and sustainable world community.

CONTENTS

INTRODUCTION

This book contains 160 no/low-cost green philanthropy ideas for families, individuals, and groups. You can choose to begin in your own home or in your own backyard or with gifts, actions, activities and projects that involve community or global outreach.

Wherever you decide to start, there are suggestions for green philanthropy appropriate for everyone in your family or group. Some are specifically geared toward youth or adults, but most can involve all ages. We hope that one or more of them will become family or group favorites that you will want to continue for years to come.

Please note that the information from websites about programs, sponsorships, adoptions, gifts, etc. was the latest available to us shortly before our book went to print. Check each website for any changes that may have been made since then.

ABOUT GREEN PHILANTHROPY

In our book we define green philanthropy as gifts, actions, activities and projects that contribute toward the conservation of all life on the planet.

A single random act of green philanthropy may not seem significant alone, but when it's multiplied tens of thousands or a million times over it brings big results, saving: small farms, urban habitats, parks and refuges, organic foods, community gardens, native and heirloom seeds, wild places, wildlife, the ancient mountains of Appalachia, forests and rainforests, rivers, lakes, and streams, and more.

In *A Sand County Almanac,* the late conservationist and philosopher Aldo Leopold wrote about a "land ethic." He saw the natural world as "a community to which we belong." He also wrote:

> The land ethic simply enlarges the boundaries of the community to include soils, waters, plants, and animals, or collectively: the land . . . That land is a community is the basic concept of ecology, but that land is to be loved and respected is an extension of ethics.

Green philanthropy means loving and respecting the community to which we belong and protecting and conserving it for future generations. As the environmentalist and writer Baba Dioum has stated:

> In the end we will conserve only what we love. We will love only what we understand. We will understand only what we are taught.

FIVE THINGS TO THINK ABOUT DOING:

- ## Create a Green Philanthropy Plan

One simple way for a family or a group to do this is to have everyone write down on separate slips of paper one or two of their favorite gifts, projects, and actions from this book and other resources. Parents or older siblings can record younger children's choices for them after you read and discuss a number of ideas with them. Put them all in a hat and draw one at a time.

Write them in the order drawn and in the appropriate section on the page provided at the end of this book: Green Philanthropy Plan. You can also download and print a planning sheet by clicking on "Create a Green Philanthropy Plan:"

www.greenphilanthropyforfamilies.org

Begin saving or raising money for the gift, project, or action that requires funds; meanwhile, start on one or two others that don't.

- ## Keep a Green Philanthropy Notebook

Encourage everyone in your family or group to write something about each gift, project, or action. Parents and older siblings can record younger children's observations, anecdotes, and stories as they relate them to you.

Include photographs and/or your own illustrations; educational pamphlets, brochures, flyers and other literature made by your family or group or collected from local and global organizations, especially if you plan to repeat a particular green philanthropy idea sometime in the future; press releases and articles; certificates or awards earned. Include the date for each notebook entry.

- ## Start a Green Philanthropy Bank

Recycle an empty jar to collect loose change in for your green philanthropy. Make your own label for it, or you can print one of ours at our website:

www.greenphilanthropyforfamilies.org

Under "Green Philanthropy Bank" click on "Download Label."

When your bank is full or your target date for accumulating funds for a gift or project has arrived, take it to a local bank that has a coin counting machine. Send a regular check or a bank cashier's check directly to the group or organization you have chosen to support, or use the money to fund a local green philanthropy project or gift.

- ## Form a Green Philanthropy Group

A group can include two or three families with children or the members of a family nature club or simplicity circle, for example. Green philanthropy groups can also be formed in faith communities, schools, environmental clubs, and other organizations.

Keep the group fairly small: twelve to fifteen members maximum and, when possible, include people of all ages. Goals might include: information sharing, fundraising, individual and group green philanthropy and community service.

Suggestions for Your Group:

1. Create a Green Philanthropy Plan.
2. Keep a *Green Philanthropy Notebook.*
3. Decide whether you want to require membership dues to fund your group's green philanthropy, hold fundraisers, or both. See Chapter Twelve, "Green Philanthropy Fundraisers."
4. Share information about one or more of your group's green philanthropy gifts, projects or actions at our website:

www.greenphilanthropyforfamilies.org

- **Share Green Philanthropy Ideas**

We encourage families, individuals, groups and organizations to submit an idea for a no/low-cost gift, project, or action for our website. It can be local, national, or global green philanthropy. Each month we'll feature selected ideas.

You can submit an idea and upload photographs on our website. Click on "Share:"

www.greenphilanthropyforfamilies.org

Chapter One: Reduce, Reuse, Recycle & Repair

> "If everybody consumed at U.S. rates, we would need 3 to 5 planets."
>
> —Facts from *The Story of Stuff*

1. **Have a family Reduce, Reuse & Recycle Scavenger Hunt.**

How many things can your family find sitting in a drawer or closet or gathering dust on a shelf that are reusable? You may want to pair up, a parent or an older sibling with a younger child, especially.

Plan a reward for after the scavenger hunt: a special green outing, for instance. Give each person the list of things to look for within an hour or whatever time limit you set if any, and then go to it: Collect things in your home, tool shed, vehicle, garage & yard, checking off the kinds of items you find and the number if your family would like to know that.

When you finish, sort through all the items first to see if there are any you can still use but have just forgotten about, and then sell the rest at a garage sale to raise money for green philanthropy. There are other things you can do with the stuff from your scavenger hunt that are included in our book.

Reduce, Reuse, and Recycle are three of the easiest things your family can do to curb global warming and save natural resources. Recycling just one aluminum can conserves enough energy to run a TV for three hours. Imagine the amount of energy, water, and other natural resources your family will be conserving when all the things you donate from your scavenger hunt get reused.

Reusable Stuff To Look For On Your Scavenger Hunt

books	____
DVDs	____
board games	____
old greeting cards	____
used cell phones	____
computer games	____
articles of clothing	____
pairs of shoes	____
eyeglasses & sunglasses	____
extra construction tools	____
extra garden tools	____
full cans of paint	
(no more than 2 years old)	____
building materials	
(lumber 6' or longer & ½ or	____
whole sheets of plywood)	____
other stuff:	____
total:	____

2. **Watch and discuss Annie Leonard's free video, *The Story of Stuff*.**

After watching it, start an ongoing list and post it on your refrigerator or bulletin board of the actions your family can take to reduce, reuse, and recycle. Share the link with other families: **www.storyofstuff.com**

Another resource, produced in partnership with nine organizations, is the following Worldwatch Institute publication: *Good Stuff? A Behind-the-Scenes Guide to the Things We Buy.* Choose one thing at a time to read about, discuss, and take action on, for example: chocolate, coffee, beverages, cell phones, and personal care products. At **www.worldwatch.org** click on "Publications" to download and save a free copy.

3. **Give your family a copy of Annie Leonard's book, *The Story of Stuff*.**

Annie Leonard's video, *The Story of Stuff,* has been viewed over ten million times on her website: **www.storyofstuff.com**. Now we have the opportunity to learn even more from her about the social and environmental impacts of our nation's over-consumption of stuff.

Your family will also discover what actions you can take in your daily lives to help build more socially just and environmentally sustainable communities that so many of us desire. When you finish reading it, pass it on.

4. Download and print "The New American Dream Wallet Buddy" for your family and carry it with you when you go shopping.

The Wallet Buddy is a simple tool to remind consumers to buy wisely, in ways that help conserve the Earth.

www.newdream.org/walletbuddy.pdf

5. Swap stuff.

Exchange extra construction and garden tools, books, DVDs, CDs, clothing, electronic and board games and additional reusable stuff with other families. A good time to swap stuff in your community is during your annual neighborhood cleanup day or on Earth Day.

Here are 6 swap sites:

www.freecycle.org
www.swaptree.com
www.swapbabygoods.com
www.swapstyle.com
www.bookins.com
www.titletrader.com

6. Donate your extra stuff to community groups that need it.

Homeless shelters, local chapters of Big Brothers Big Sisters, and domestic violence shelters for women and children are among the many nonprofit organizations to consider. Call first and ask about their current needs.

7. Sell your unwanted stuff through eBay's Giving Works program and donate the profits to your favorite environmental group.

Your family or a group you belong to can sell unwanted items through the Giving Works program and donate a portion or all of the funds raised to the local or national organization of your choice. Over 17,000 nonprofits of every size are registered, with nearly 150 million dollars having been raised shortly before this book went to print.

www.ebaygivingworks.com/nonprofit-info.html

A nonprofit called Mission Fish: **www.missionfish.org** screens the organizations in the Giving Works program. If your favorite local or national environmental organization isn't listed they can register for free at:

www.missionfish.org/charity/web.us/ npregrecommend1.jsp

8. Recycle your gently used books into reading gift packs for friends.

This is a fun way to exchange books. A youth gift pack might include: two or three books; a handmade or recycled postcard* inside one or more of the books with a brief description of what you enjoyed most about it; an extra bookmark you don't use or a handmade one; a reading game or activity from the Reading is Fundamental website printed on recycled paper:

www.rif.org

Tie each gift pack together with ribbon or twine.

*See #16: "Make postcards and bookmarks from old greeting cards as gifts."

9. Donate gently used books to these places.

- Many Reach Out and Read programs accept gently used books. Contact a Reach Out and Read site in your community to learn more about their specific needs:

www.reachoutandread.org

- Literacy programs. To find locations in your community, visit:

http://www.literacydirectory.org

- Most public libraries accept gifts of gently used books. They may either add them to their collection or include them in a book sale to raise money for the library. Call your neighborhood library for specific information about donating.

- Boys Booked on Barbershops (B-Bob) & Girls Booked on Beauty Shops (G-Bob):

www.mochamoms.org

- Homeless Shelters: Many shelters that serve families accept books for children of all ages. Call first and ask about their specific needs. Check the following online directory to locate homeless shelters in your area:

www.homelessshelterdirectory.org

<u>About Reach Out and Read:</u>

This nonprofit organization provides information and training for pediatricians, nurses, residents and other health professionals on the importance of reading early to children. This information is then shared with parents of children between the ages of 6 months to 5 years during regular checkups. With each visit a child receives a developmentally appropriate book to take home and keep. The focus is on kids growing up in poverty.

There are Reach Out and Read sites in all 50 states, located in hospitals, health centers, clinics and private doctors' offices—about 3800 in all.

To find a program in your area, go to:

www.reachoutandread.org

10. Donate your used eyeglasses to the Lions Recycle for Sight Program.

Both prescription and non-prescription eyeglasses and sunglasses with either metal or plastic frames are accepted. To find a collection site near you, call a local Lions Club or visit their website: **www.lionsclubs.org** and click on "Donate Eyeglasses."

11. Buy electronics from companies that will take back (recycle) your old products.

The Electronics Take Back Coalition provides a list of electronics companies that will take back their products either for free or a charge:

www.electronicstakeback.com

12. Recycle your old electronics with the most socially and environmentally responsible e-recyclers.

At the website: **www.e-stewards.org** click on "Find Your Local E-Stewards" on the menu bar. Share the link with others.

Check out Annie Leonard's video: *The Story of Electronics:*

www.storyofstuff.com

13. Donate building materials and supplies left over from your home fix-up projects.

Your local Habitat for Humanity ReStore will either sell them to raise money to build more houses or will give them directly to a building or remodeling project in your community.

Visit **www.habitat.org/env/restores.aspx** to find a ReStore in your community. Call and ask for a list of needed items.

Rebuilding Together is another organization that accepts donations of building materials and supplies.

www.rebuildingtogether.org

14. Donate extra construction and garden tools.

Donate extra construction tools around your house, used or new, to your local Rebuilding Together affiliate or to the Habitat for Humanity ReStore. Garden and lawn care tools may also be needed.

www.rebuildingtogether.org
www.habitat.org/cd/local/

A local community garden is another place that may need garden tools. At the American Community Gardening Association website you can search for a community garden by name or by zip code, city or state:

www.communitygarden.org

At this same website by clicking on "Learn" and then "Best Practices" you'll find the tip sheet: "Ten Tools Every Community Gardener and Garden Needs."

15. Give used greeting cards to St. Jude's Ranch for Children.

The St. Jude's Ranch for Children accepts donations of all-occasion greeting cards. The youth help make new cards, replacing the backs with recycled cardstock. They're sold online and in their Gift Shop and Resale Boutique, with all proceeds benefiting the children at all three ranches. Mail card fronts only to:

St. Jude's Ranch for Children, 100 St. Jude's St., Boulder City, NV 89005-1681.

www.stjudesranch.org

16. Make postcards & bookmarks from old greeting cards as gifts.

This simple family project saves trees—and money. Postcard sizes range from 3-1/2" by 5" to 4-1/2" x 6". Anything larger requires more postage. Cut templates out of cardboard from your recycling bin to use as guides.

17. Have a make-your-own-greeting-cards party.

Invite family and friends, or the members of your ecology club, simplicity circle, or another group to which your family belongs. Ask everyone to bring any extra envelopes they have and small objects, for example: pressed and dried flowers, seeds, different colors and textures of paper and light cardboard from their recycling bins, wallpaper samples, fabric scraps, and artwork. Sheets of cardstock and blank, pre-folded cards with matching envelopes are available at most hobby and craft stores.

This project can be used as a group fundraiser.

18. Choose a project from the *Make a Difference Middle School Kit* for your family to work on together.

The kit is one of many free environmental education resources available from the EPA. Some materials in it are downloadable. Even though it was written for middle school youth, families and groups can use parts of it with kids of all ages.

Choose a project that your family can work on together to reduce waste and conserve resources in your home, school, or community.

Here are some of the booklets and posters included in the kit:

- "Greenscaping" Your Lawn and Garden
- Let's Go Green Shopping
- Make a Difference in Your School: A How-to Guide for Engaging Students in Resource Conservation and Waste Reduction
- The New Wave in Electronics: eCycling
- Science Fair Fun
- Service Learning: Education Beyond the Classroom
- Tools to Reduce Waste in Schools
- The Life Cycle of a CD or DVD
- The Life Cycle of a Cell Phone
- The Life Cycle of a Soccer Ball
- Pack a Waste Free Lunch

To order a kit (only one per household) log onto:

www.epa.gov/osw/education/ordermad-ms.htm

More Reduce, Reuse, Recycle Resources

The free 44-page guide, *Tools to Reduce Waste in Schools* is a great resource for high school students interested in starting a reduce, reuse and recycle program. To download a copy, log onto:

www.epa.gov/epawaste/education/toolkit.htm

Check out EPA's High School Environmental Center for more resources: **www.epa.gov/highschool**

America Recycles Day.Org also provides free resources, including: posters, flyers, bookmarks, the *Little Green Activities Book,* and more:

www.americarecyclesday.org/arddownloads.aspx

Resources on Zero Waste events:

www.ecocycle.org/zwevents/

The Grassroots Recycling Network provides resources for Zero Waste events:

www.grrn.org/zerowaste/index.html

19. Donate a book about the 3 R's to your neighborhood school or public library.

Six gift ideas:

- *Recycle: A Handbook for Kids,* by Gail Gibbons
- *The Adventures of an Aluminum Can: A Story About Recycling,* by Alison Inches

- *The Adventures of a Plastic Bottle: A Story About Recycling,* by Alison Inches
- *The Three R's: Reuse, Reduce, Recycle,* by Nuria Roca
- *Where Does the Garbage Go?* by Paul Showers
- *Why Should I Recycle?* by Jen Green

20. Make green philanthropy wish lists for birthdays and holidays.

Instead of getting more stuff, everyone in your family could be given the option of putting the money toward their favorite green philanthropy gift or project. They might choose, for example to: plant so many trees through American Forests or Nature Conservancy; adopt an endangered wildlife animal; make a contribution to help save BioGems. There are many more gift ideas in this book and other resources.

21. Add another "r" to reduce, reuse, and recycle: repair.

Do an Internet search on how to repair things and millions of results pop up. If everyone, including older youth, learned how to repair just one thing, a lot of natural resources and landfill space would be saved.

Some of the simple things youth can do include: learn to sew on a button, mend a tear in an article of clothing, glue a broken part back on a wooden toy or piece of furniture, and learn other repair skills by helping a parent or grandparent.

As Julia Butterfly Hill and other conservationists remind us, when we say that we are going to throw something away, we need to stop and ask ourselves, "Where is *away*? There is no such thing."

On the following website your family can learn to repair things and share your own how-to information with others:

www.ehow.com

Chapter Two: Simple Living

"As an end, the adoption of simple lifestyles by Western megaconsumers is truly the only way we're going to save the planet."

—Marie Sherlock, author of
Living Simply With Children

22. Have a Simplicity Day often.

Seven Suggestions

- Sometime before your simplicity day, have a Reduce, Reuse & Recycle Scavenger Hunt and give away what you don't need or use.
- On your simplicity day spend time, individually or as a family, doing something that's important to you.
- Limit the time you spend on email, the Internet and television in order to free up time to do what you love.
- Read, walk, garden, write in your journal, do yoga or some other kind of activity that inspires meditation and creative thinking.
- Buy nothing or as little as possible; get by with what you have.
- Learn to simplify household tasks so that you can spend time on what's important to you.

- Write an affirmation for your simplicity day, either individually or as a family, for example: “I live a simple and peaceful life.” “I bring peace and happiness into my life when I do what I love and say no to the rest.” Repeat it to yourself often throughout the day.

23. Simplify the Holidays.

Six Things Your Family Can Do

- Give one another less stuff and more time.
- Spend green hours together, focusing on the things that money can’t buy.
- Work on a green philanthropy project together.
- Give an alternative gift to an individual or organization in your community.
- Gather up things you no longer need or use and donate them to a local organization.
- Use less stuff: use your own dishes instead of disposables; create fun ways to package gifts; make your own greeting cards.

24. Join a simplicity study group.

The Simple Living Network’s website maintains a database of local simplicity groups. Click on “Study Groups & Circles” at:

www.simpleliving.net

25. Start your own simplicity study group.

This can be part of an environmental club or sustainability program, for instance, or a separate group. *The Circle of Simplicity: Return to the Good Life,* by Cecile Andrews provides a format for creating a group. Information about the book, including group discounts, is available at:

www.simpleliving.net

Another book we recommend: *Living Simply With Children: A Voluntary Simplicity Guide for Moms, Dads, and Kids Who Want to Reclaim the Bliss of Childhood and the Joy of Parenting,* by Marie Sherlock

Chapter Three: Community Service

"Nobody can do everything, but everyone can do something."

—Author Unknown

26. Do a family green philanthropy project in your community for Earth Day or any day of the year.

A project doesn't have to be complicated or time-consuming. At Serve.Gov your family can search for local environmental community service projects. You'll also find toolkits on this same website for creating your own project and registering it:

www.serve.gov

The Earth Day Network website is another place to register your family project and to involve other volunteers in your community if you need them. The toolkit, *Earth Day in a Box* provides information on planning and sponsoring a project:

www.earthday.net

27. Youth: Find an environmental community service project for National Youth Service Day that you and your family can work on together.

The Youth Service America website, geared toward youth ages 5 to 25, provides the tools, information and resources to plan, design and complete a service project locally, nationally or globally. Resources include:

- Project Plan-It, an online interactive tool that enables youth to plan, design, implement and manage a service project
- *National & Global Youth Service-Learning Curriculum Guide*

For more information, visit: **www.ysa.org**

Global Youth Service Day: **www.gysd.org**

If you would like to receive information about grants by email, subscribe to YSA's Service Wire:

www.servicewire.org/nsb/Grants-and-Awards

Resources

The following book includes environmental community service projects:

The Kid's Guide to Service Projects: Over 500 Service Ideas for Young People Who Want to Make a Difference, by Barbara A. Lewis

You can also download and save free EPA publications about community service projects including the following:

Environmental Protection Begins With You, Linking Girls to the Land, Service-Learning Beyond the Classroom, and *Volunteer for Change:*

www.epa.gov/teachers/community-svc-projects.htm

28. Collect gently used books for a group that needs them.

Suggestions for places to collect them include: grocery stores, parents and grandparents' workplaces, and at a table or booth during your local Earth Day event. Make posters to advertise your event and write a notice or a press release for your local newspaper.

See # 9: "Donate gently used books to these places."

Three ways to make used books look almost new:

- Hardback books generally have an extra blank page before the title page. If a personal message was written on the inside cover of the book, simply glue the extra page over it.
- If the personal message is small, go to the website: **www.myhomelibrary.org** and print a bookplate big enough to glue over it.
- Glossy book covers can be wiped clean with a damp cloth.

29. Adopt an under-resourced classroom or school library in your community.

Sponsor a book-recycling day in your community for your

adopted school or library. Put collection boxes and posters in grocery stores and other businesses.

Nationwide surveys show that teachers spend hundreds of dollars of their own money each year on educational materials and supplies for their classrooms, including books, which can add up quickly.

30. Collect used eyeglasses for the Lions Recycle for Sight Program.

According to the World Health Organization 153 million people worldwide need glasses. Ask the principal of your neighborhood school to sponsor the program. Or, if your family belongs to an environmental club or organization that participates in your area's annual Earth Day event, propose the project for your table or booth.

This can be a separate community service project or part of a larger reduce, reuse, and recycle Earth Day campaign. To find a drop-off collection site in your area, log onto **www.lionsclubs.org** and click on "Donate Eyeglasses."

31. Sponsor an education campaign about recycling old electronics in your community.

This can be part of a general education campaign about reducing, reusing, and recycling, or a separate community service project. Make posters about recycling old electronics and distribute them to local grocery stores and other businesses, religious organizations, schools, and libraries. Ask neighborhood associations to publish information about recycling electronics in their newsletters with locations of e-stewards in your community and their hours of operation.

Good times to sponsor this event include: America Recycles Day, **www.americarecyclesday.org**, your area's annual Earth Day event: **www.earthday.net**, and during neighborhood associations' yearly cleanups.

Resources for Your Education Campaign

The Electronics Take Back Coalition:

www.electronicstakeback.com

To find local e-stewards: **www.e-stewards.org**

The Silicon Valley Toxics Coalition website has the *Electronics Purchasing Guide* and the *Electronics Recycling Guide* in addition to other resources: **www.etoxics.org**

The Story of Electronics: **www.storyofstuff.com**

32. Volunteer in or help start a community garden.

At the website of the American Community Gardening Association, your family can search for community gardens near you by the name of the garden or by zip code, city or state. On the main web page click "Connect:"

www.communitygarden.org

At this same website you'll find numerous resources for starting a community garden including funding opportunities. Check out the series of downloadable "Best Practices Tip Sheets" collected from community gardens all across the country. These include: *10 Tips For Starting A*

Community Garden, 10 Kids Gardening Tips, Tips For Starting Seeds, and *How To Save Seeds.*

33. Help start a school garden.

To learn how to plan and create a school garden, download and save the free 51-page, *Getting Started: A Guide for Creating School Gardens as Outdoor Classrooms.* Published by the Center for Ecoliteracy in collaboration with the Life Lab Science Program, it's available at:

www.ecoliteracy.org/publications/getting-started.html

The Edible Schoolyard and the Chez Panisse Foundation publish the following books: *Ten Years of Education at the Edible Schoolyard* and *Principles of an Edible Education: A Vision for School Lunch*:

www.edibleschoolyard.org
www.chezpanissefoundation.org

Sowing the Seeds of Success: How to Start and Sustain a Kids' Gardening Project in Your Community is one of many resources available at: **www.kidsgardening.org**

You can also pick up a lot of useful information and advice from the directors of already-established school and community gardens in your area. One or more of them may have a mentoring program in which older youth and adults help other groups get started.

Grants/Funding For A School Garden

Environmental Protection Agency:

www.epa.gov/teachers/grants.htm

The KidsGardening Resource Directory provides links to grants and funding as well as free materials and other educational information:

www.kidsgardening.com/resources/resource.asp

National Environmental Education Foundation:

www.neefusa.org

National Wildlife Federation:

www.nwf.org/schoolyardhabitats

34. Parents and Older Youth: Start a Junior Master Garden Club.

It doesn't cost anything to become a JMG Club, and the recommended minimum number of members is only five. The leader is not required to be a Master Gardener, either.

Service learning is a part of the program: Older youth who have earned certification as Junior Master Gardeners are encouraged to become mentors to younger children.

The Level 1 Junior Master Gardener program is designed for grades 3-5 and Level 2 is for grades 6-8. For more information, visit: **www.jmgkids.us**

In addition to their program guides, JMG publications include: *Wildlife Gardener* and *Literature in the Garden.*

35. Provide plants for local Habitat for Humanity homeowners or a community garden.

To download Habitat for Humanity's step-by-step pdf guide, *Seeds of the Future,* log onto:

http://www.habitat.org/youthprograms/

36. Protect and conserve the groundwater in your area.

Groundwater.Org lists a number of community service projects youth and their families can do. These include:

- Pick up litter in your neighborhood and on your school grounds.
- Talk with your school's principal and superintendent about certification as a Groundwater Guardian Green Site. Step-by-step instructions are included.
- Ask friends or family members to help stencil a groundwater friendly message on storm drains in your area. Instructions are provided, including where to get permission.
- Make posters or flyers to educate others on ways to conserve water.
- Talk with your teacher about a groundwater class project or a school-wide groundwater education day.

www.groundwater.org/ta/kidsprotect.html

37. Volunteer to help with a local watershed project.

The website **www.epa.gov/adopt** maintains a database of 2600 watershed groups, searchable by zip code, nationwide. Your family can volunteer for a project with a local group or plan your own using the EPA's *Watershed Stewardship Toolkit.*

Examples of watershed projects include: putting door hangers on houses to remind people not to dump hazardous waste in storm drains, monitoring a stream's water quality, and stenciling reminders on storm drains.

Your family can also register your watershed project at **www.serve.gov** and request other volunteers if you need them.

Additional resources to help you create your own watershed project:

Resources for watershed education and action:

www.epa.gov/adopt/resources/

Give Water a Hand Action Guide: Free downloadable guide at the website of the University of Wisconsin Extension Office: **www.uwex.edu/erc/gwah/**

Girl Scouts' Water Drop Patch Project:

www.epa.gov/adopt/patch

EPA Watershed Patch Project:

www.epa.gov/adopt/patch/certificates

38. Sponsor an education campaign about household hazardous waste.

Annual Earth Day events are a good time for families, eco clubs, sustainability groups and other environmental organizations to do a one-day community service project to educate others on this issue. Provide information about where they can take their used motor oil, toxic cleaners, old cans of paint and other household hazardous waste to be disposed of properly; include the facility's hours of operation. To search for services in your area, log onto:

www.earth911.com

Contact them about your project and ask if they can provide educational materials such as door hangers, flyers, and brochures. Post information about your project on community bulletin boards at grocery stores and other retail businesses, libraries, and other places. Send or email a press release about your event to your local newspaper.

39. Adopt a park.

Adopt a nearby park or part of a trail and help keep it clean by picking up litter. Carry a couple of bags on your walks or picnics, one for trash and the other for recycleables.

40. Help create a Schoolyard Habitat.

The Schoolyard Habitat is a National Wildlife Federation program. It's an outdoor classroom in which youth have opportunities to learn firsthand about ecosystems, wildlife species, and conservation. Cross-curricular studies enhance the outdoor learning experience. Online resources include: ideas about where to get help in starting a Schoolyard Habitat, lesson plans, and other curriculum resources.

These two articles are also posted on the website: "The Schoolyard Habitat Movement: What It Is and Why Children Need It," by Mary Rivkin and "Down By the Schoolyard," by Stephen R. Coffee:

www.nwf.org/schoolyard/

41. Volunteer to help remove invasive plants in a nearby park or other public land.

Search for volunteer opportunities in your area at these websites:

National Public Lands Day – a National Environmental Education Foundation Program:

www.publiclandsday.org/managers/invasives.htm

Serve.Gov: **www.serve.gov**

<u>To learn more about invasive plants check out these resources:</u>

The National Parks Service provides illustrated and easy-to-read fact sheets on invasive plants at this website:

www.nps.gov/plants/alien/factmain.htm

"Aliens in Your Neighborhood" is another project of the National Parks Service:

www.nps.gov/invspcurr/alienhome.htm

It includes curricula for classroom instruction and field excursions to parks. The curricula can be adapted to home use.

The United States National Arboretum provides links to state information sites:

www.usna.usda.gov

The Department of Agriculture's National Invasive Species Information Center:

www.invasivespeciesinfo.gov

42. Adopt a local wildlife refuge.

There are more than 530 National Wildlife Refuges in the United States, at least one in every state. These refuges protect habitat for migratory birds, endangered species and other wildlife. To find refuges near you, visit the website of the U.S. Fish and Wildlife Service:

www.fws.gov/refuges

Examples of wildlife refuge service projects for youth and their families include:

- Pick up trash and litter
- Build birdhouses
- Help biologists count wildlife

- Plant habitat for wildlife
- Read and learn about wildlife, their habitats, and about wildlife refuges; share that information with others

43. Create a Junk Mail Monster for Earth Day.

This is a great way to raise awareness about junk mail and how many millions of trees are cut down every year to produce it. The Junk Mail Monster can be a real person or an effigy. Glue junk mail in layers to old clothing, including a large hat, until it's completely covered.

The Junk Mail Monster provides visitors to your area's annual Earth Day event the link to the following website written on the backs of pieces of junk mail:

www.donotmail.org/form.php?id=50

Search "Junk Mail Monsters" online to see how different groups have depicted theirs.

Chapter Four: Ecoliteracy

> "In the end we will conserve only what we love. We will love only what we understand. We will understand only what we are taught."
>
> —Baba Dioum

44. Give your family a "Green Hour" often.

Research shows that spending time engaged in nature experiences and outdoor play improves intellectual and creative thinking, emotional wellbeing, self-discipline and physical fitness. It also lowers stress levels and reduces symptoms of ADD and ADHD.

Green hours also nurture and grow future environmentalists and conservationists. Studies show that the one factor people most often credit for having had the greatest influence on their attitude toward the environment and conservation was their direct experience with nature while growing up: **www.greenhour.org**

Resources for fun and educational ways to spend green time with family and friends:

Green Hour E-Newsletter:

www.greenhour.org/section/about/contact/e_newsletter

Nature Explore provides resources for families and educators, including instructions for creating outdoor play spaces: **www.arborday.org/explore/**

To find nature nearby, go to: **www.nwf.org/naturefind/**

For more outdoor activities, visit:

www.learnoutside.org/familyactivities.html

National Wildlife Federation's *Be Out There* initiative:

www.nwf.org/beoutthere/

Children and Nature Network founded by Richard Louv, author of, *Last Child in the Woods: Saving Our Children from Nature-Deficit Disorder*:

www.childrenandnature.org

Free downloadable booklet: *A Parents' Guide to Nature Play,* by Ken Finch, founder of Green Hearts Institute:

www.greenheartsinc.org

45. Parents: Learn about nature deficit disorder and ways to prevent it.

A series of videos called *EcoSense for Living* focuses on nature deficit disorder and what parents can do to prevent it in their own children. To view them, go to **www.childrenandnature.org** and enter "EcoSense for Living Videos" in the search box. Share the link with other families.

46. Read and discuss the book, *Smart by Nature: Schooling for Sustainability.*

Written by Michael K. Stone of the Center for Ecoliteracy, this is one of the best books available on the topic. Your family will learn about the growing sustainable living movement in K-12 schools across the country and the latest innovative ideas and hands-on experiences and projects developed from it. When you finish reading it, pass it on to a teacher, friend, neighbor or sustainability group.

If your neighborhood school doesn't already have a sustainable living program, discuss the idea with the principal.

47. Parents: Give your child's teacher a subscription to *Green Teacher.*

Each issue of the *Green Teacher* magazine, published four times a year, has fifty pages of environmental ideas, activities and resources for youth ages 6 to 18. To download a free sample issue, log onto:

www.greenteacher.com

48. Give a family with young children (your own?) a book about nature play.

These two provide many fun ways to enjoy green hours together:

I Love Dirt! – 52 Activities to Help You and Your Kids Discover the Wonders of Nature, by Jennifer Ward. Ages 4-9.

Nature's Playground: Activities, Crafts, and Games to Encourage Children to Get Outdoors, by Fiona Danks and Jo Schofield. Ages 7-9.

49. Start a nature club for families.

To download a free online toolkit for starting and maintaining a club, log onto **www.childrenandnature.org** and click on "Nature Clubs for Families."

50. Make a nature journal for each member of your family.

Put together your own journals or download a cover template at the Sierra Club's Environmental Education website:

www.sierraclub.org/education/nature_journal.asp

A nature journal is "a place to grow your thoughts, feelings, ideas, activities, observations, and relationship with the natural world." Parents or older siblings can write what younger children dictate to them. Illustrate your journals, if you want, with photographs, pencil drawings, crayons, paints or magic markers.

At this same website you can also download the free *John Muir Day Study Guide.* It was developed to help teachers and students commemorate John Muir Day on April 21st, but the lesson plans: about plant and animal habitats, ecosystems, biodiversity, forests and other topics can be used at home, too.

51. Start a nature book club.

Alternatively, if you already belong to a book club, propose the idea of adding books about nature and the environment to the reading list. Three recommendations for adults:

Coyote's Guide to Connecting with Nature, by Jon Young, Ellen Haas, and Evan McGown

Last Child in the Woods: Saving Our Children from Nature-Deficit Disorder, by Richard Louv

The Geography of Childhood: Why Children Need Wild Places, by Gary Paul Nabhan and Stephen Trimble

Youth: If your school doesn't already have a nature book club, talk with your teacher about starting one in your school or classroom, eco club, or sustainability program.

A Resource for Youth Nature Book Clubs

A great downloadable resource to help your club put together a reading list is on Project Learning Tree's website: **www.plt.org**

Click on "Curriculum" in the menu bar on the left and then "PreK-8 Guide." Scroll down to "Children's literature list—sorted by author or by PLT activity."

A number of books on the list are out-of-print but may still be available at your local school or public library, a used bookstore, or on Amazon. Schools all across the country participate in Project Learning Tree's award-winning environmental education program. Ask your teacher if your school is one of them.

52. Ask the principal if your school can participate in the *Green Schools! Initiative*.

The *Green Schools! Initiative* includes these five areas of study and action: energy, water, waste and recycling, school site, and environmental quality. For more information, visit:

www.forestfoundation.org/cel_greenschools.html

Chapter Five: In Your Own Backyard

"A lawn is nature under totalitarian rule."

—Michael Pollan

53. Give your family a healthy yard.

This project benefits your family, pets, and the environment. Two of the best resources we've found are:

Noah's Garden: Restoring the Ecology of Our Own Backyards, by Sara B. Stein

The Organic Lawn Care Manual: A Natural, Low-Maintenance System for a Beautiful, Safe Lawn, by Paul Tukey, founder of Safe Lawns.Org

You'll find how-to videos at **www.safelawns.org** including:

Why Organic, Why Now?
What's in the Organic Bag?
Aeration

Additional resources for creating a biodiverse yard are available for free at **www.beyondpesticides.org**

54. Donate a copy of *Noah's Garden* or *The Organic Lawn Care Manual* to your neighborhood public library.

55. Invite an expert to give a how-to workshop on organic lawn care.

Family nature clubs, neighborhood associations, eco clubs, and faith-based Earth stewardship programs are just some of the groups that may want to consider sponsoring a workshop on this topic. Check the yellow pages of your telephone book for organic lawn care businesses, or call a local nursery that emphasizes organic lawn care.

Alternatively, your group can sponsor a how-to workshop using *Noah's Garden: Restoring the Ecology of Our Own Backyards, The Organic Lawn Care Manual,* how-to videos at Safe Lawns.Org and other resources.

56. Put a warning decal on the curb above the storm drain near your home.

This is an easier alternative to stenciling, although the decals don't last as long.

Whatever we put down storm drains flows directly into our lakes and streams. It isn't run through our sanitary sewer plants, and so it doesn't receive any kind of treatment. Two things your family can do to help keep the lakes and streams in your area clean and healthy are:

- Take used motor oil, antifreeze, paint, toxic cleaners and other household hazardous waste to a facility that responsibly handles it. To find one in your area, go to: **www.earth911.com**

- Print an EPA storm water decal from the website below and have it laminated. Using cement glue attach it to the curb above the storm drain that's near your home. There are twelve different designs to choose from at:

www.epa.gov/npdes/pubs/stormwaterstickers.pdf

57. Grow native plants.

A fun way for your family to learn about native plants is to visit a local nursery that emphasizes landscaping with them. Talk with the people there and check out what's available. Write down the names of recommended plants and information about them in a notebook so that you can later go online and learn more about each one.

Your family may want to work out a plan together on paper first. Give everyone the opportunity to select at least one or two of their favorite plants. These websites provide more information about landscaping with native species and why they are beneficial to local ecosystems:

Lady Bird Johnson Wildflower Center:

www.wildflower.org

North American Native Plant Society: **www.nanps.org**

Wild Ones: **www.for-wild.org**

How to Make Your Own Biodegradable Seedling Pots

Cut strips of newspaper about 13" long and 5" wide. Using a soup can or a larger size if needed, place it on two layers of strips, leaving a couple inches of newspaper below the can. Roll into a cylinder. Push together the newspaper underneath it to shape it into a pot. Set it down in a box, remove the can, and fill with potting soil.

58. Exchange native plants with neighbors & friends.

Organize an exchange of native plants on your block, in your neighborhood association, community garden, religious organization, or ecology club. Invite an expert from your local native plant society or from a nursery that emphasizes native plants to talk with your group about landscaping with them and how they benefit the local ecosystem.

59. Plant a native tree to celebrate a special day.

Celebrate a birthday, anniversary, Earth Day, or another special day by giving someone in your extended family a gift certificate to a local nursery that maintains a good selection of native trees and bushes.

For step-by-step instructions on how to plant a tree, go to:

www.americanforests.org/planttrees/howto.php

60. Create a Nature Explore play space in your back-yard.

The Arbor Day Foundation's *Learning With Nature Idea*

Book has step-by-step instructions for designing and creating an outdoor play space in your own backyard, a school, nature center, and other sites. There are plans to fit all budgets. Your family can take a virtual tour of an outdoor Nature Explore play space at this web address:

www.arborday.org/explore/

Another great resource:

Natural Playscapes: Creating Outdoor Play Environments for the Soul, by Rusty Keeler.

61. Give your family nature-exploring kits.

Use them in your own backyard or take them along on nature outings. They can be tailored to everyone's individual interests.

Things to include in a kit: a journal or notebook, binoculars, magnifying glass, camera, flashlight for exploring at night, and age-appropriate nature guides to birds, butterflies, backyard wildlife, insects, wildflowers, etc.

These also make great gifts for friends who enjoy exploring nature.

62. Create a wildlife habitat in your yard and get it certified.

The National Wildlife Federation website provides step-by-step instructions on how to create a backyard habitat and have it certified: **www.nwf.org/backyard/**

63. Conserve the wildlife in your own backyard.

The first step to doing that is to learn about the different wildlife species in your backyard habitat. Here are four gift ideas we recommend:

Attracting Birds, Butterflies and Other Backyard Wildlife, published by the National Wildlife Federation.

Backyard Wilderness, by Catherine Herbert Howell. This is one in a series of National Geographic's *My First Pocket Guide.*

Two more books in the series:
Wildflowers and *Garden Birds*

64. Create a family conservation guide to the wildlife in your own backyard.

As you learn about earthworms, bees, butterflies, birds, Ladybird beetles, raccoons, opossums, rabbits, garter snakes, bats, fireflies and other species in your backyard habitat, write about each of them in your family guide. Include photographs or your own illustrations.

A lot of free information is available online. The example below is a partial list from the Nature Conservancy website of everyday things your family can do to conserve birds. To read the complete list, go to:

www.nature.org/initiatives/programs/birds/issues

Seven things you can do to conserve birds in your own backyard:

- Put up a birdhouse with proper ventilation.
- Put a birdbath in your yard. Use a heater in cold weather.
- Erect bird feeders and nectar feeders away from windows and where they will be safe from predators.
- Limit the use of lawn chemicals and pesticides.
- Hang cut-out silhouettes of birds such as hawks in large windows to keep birds from flying into them.
- Plant native fruit and berry-bearing bushes and trees.
- Learn about the common birds in your neighborhood and the value of all wildlife.

More websites with free resources about conserving birds

Bird Education Alliance for Conservation:
www.birdedalliance.org

Cornell Lab of Ornithology: **www.birds.cornell.edu**

Flying Wild – a program of the Council for Environmental Education: **www.flyingwild.org**

International Migratory Bird Day: **www.birdday.org/**

An Audio Guide to Urban Birds was created by youth for youth. Hear recordings of birds' songs and discover tips for remembering their calls. Read descriptions of birds and take short quizzes to test your identification skills. For more information or to download a free CD and booklet, visit: **www.urbanbirdsounds.org**

65. Build a good nest box for your backyard habitat.

On the following website, scientists at the Cornell Lab of Ornithology discuss the general features of a good nest box and provide detailed instructions on how to build nest boxes for different species of birds:

www.birds.cornell.edu/nestinginfo/nestboxref/construct

66. Register your nest box and participate in NestWatch.

NestWatch is a fun and educational way for your family to learn about the lifecycles of bird families. By monitoring your nest box once or twice a week and reporting your information online to the Cornell Lab of Ornithology, you will be helping scientists study how bird families in your area are affected by climate change.

For more information, visit:

www.nestwatch.org
www.RegisterYourNestbox.org

Chapter Six: In The Garden

> "The greatest gift of the garden is the restoration of the five senses."
>
> —Hanna Rion

67. Parents: Give your kids their own garden.

Even if their families have a garden, most kids of all ages would probably enjoy planting and tending one of their own if given the opportunity. A parent or an older sibling can give a younger child this gift. And it doesn't have to be big; a 4' x 4' or 4' x 6' area or a container garden on an apartment balcony will do.

The University of Illinois Extension Program provides a free online guide called, *First Garden*:

www.urbanext.uiuc.edu/firstgarden/

The Kids Gardening website has the free downloadable *Parents' Primer* with complete step-by-step instructions, including a chapter on how to build a small garden, "Starting Small: A Place a Kid Can Call Her Own:"

www.kidsgardening.org

Rodale's Ultimate Encyclopedia of Organic Gardening provides instructions for creating different kinds of gardens, including:

Container Gardens, Raised Bed Gardens, Kitchen Gardens, Rock Gardens, Shade Gardens, and Water Gardens.

68. Give them each a garden kit.

A kit might include, among other things: child-size tools, work gloves, a copy of *My Gardening Journal*, a tote to keep the smaller tools and work gloves in, and a gift certificate to a local nursery for seeds or plants.

For instructions on making your own biodegradable seedling pots, see #57: "Grow native plants."

Gardening With Kids.Org publishes *My Gardening Journal*, just for kids: **www.gardeningwithkids.org**

If your family or a group you belong to, would like to donate garden kits to children in your community, contact the director of a local youth garden and ask about their specific needs. To find community gardens in your area, log onto: **www.communitygarden.org**

69. Give Butterfly Garden Kits as Earth Day and Mother's Day gifts.

Buy seed packets of one or more native plant species that attract butterflies and other pollinators. Check the Internet for local companies that sell native seeds wholesale or in bulk. Members of your local Native Plant Society may also be willing to donate seeds.

Type and print the following instructions for creating a butterfly garden and staple a copy to each packet of seeds.

Six Steps to Create a Butterfly Garden

- For best results choose a spot with full sun.
- Provide flat stones where they can rest and bask in the sun.
- Plant a wide range of native perennials that will provide a succession of flowers, and thus nectar and pollen, through the entire growing season. Check your local nursery for native plants that attract butterflies and other pollinators. There is also a database of native plants at:

  www.wildflower.org/alternatives/

- Learn which species of butterflies are native to your area and which plants will attract them at this website:

  www.butterfliesandmoths.org/map

  Type in the name of your county and click on "Show Location." Beautiful photographs accompany the text.
- Avoid insecticides and herbicides. Use organic lawn and garden products and natural insect and weed control methods, instead.
- Give them a muddy or moist, sandy place for puddling, where they can drink water and obtain minerals from the soil.

Resources for Youth

Nature's Partners: Pollinators, Plants, and You is a free online curriculum for grades 3 through 6 on the North American Pollinator Protection Campaign (NAPPC) website: **www.nappc.org/curriculum/**

Gardening With Kids.Org publishes the *Pollinator Field Journal:* **www.gardeningwithkids.org**

70. Build a simple composter and learn how to compost.

We recommend:

Easy Composters You Can Build,
by Nick Noyes

Another book we like is: *Let It Rot! The Gardener's Guide to Composting*,
by Stu Campbell.

When you finish either book, pass it on to another family or donate it to a community garden or your neighborhood library.

Free information on composting is available at these websites:

www.mastercomposter.com
www.howtocompost.org
www.epa.gov/compost
www.wormwoman.com

71. Save and swap seeds.

Here are several resources to help your family get started on this project:

Saving Seeds: The Gardener's Guide to Growing and Storing Vegetable and Flower Seeds, by Marc Rogers

"Seed Saving For Beginners:"

www.fedcoseeds.com/seeds/seed_saving.htm

Another resource at this same website: At the bottom of the menu bar on the left click on "Seed School Curriculum" to download, *From Generation to Generation: An Activity Guidebook in the Living Tradition of Seed Saving,* by Eli Kaufman.

Seed Savers Exchange: **www.seedsavers.org**

National Gardening Association:

www.garden.org/seedswap/

To swap with another school garden:

www.kidsgardening.com/school/searchform.asp

72. Donate money from your Green Philanthropy Bank to a local community garden.

Community gardens use donations to fund: tools, seeds, plants, and youth education among other things. To find a community garden in your area by the name of the garden or by zip, city or state, visit the website of the American Community Gardening Association:

www.communitygarden.org

73. Give a garden kit to a family in a developing country.

A Mercy Corps Family Garden Kit provides a family living in an impoverished village the means to raise their own vegetables and improve their health. Each kit includes tools, seeds, gardening instructions, and information on children's nutritional needs. For more information, log onto:

www.mercycorps.org/mercykits

74. Donate a new or gently used book about gardens or seeds for youth to your neighborhood school or public library.

Here are 11 gift ideas:

- *Gardening Wizardry for Kids*, by L. Patricia Kite and Yvette Santiago Banek
- *Grandma's Garden,* by Mercer Mayer
- *Green Thumbs: A Kids Activity Guide to Indoor and Outdoor Gardening,* by Laurie Carlson
- *How a Seed Grows*, by Helen J. Jordon
- *Jack's Garden*, by Henry Cole
- *Planting a Rainbow*, by Lois Ehlert
- *Roots, Shoots, Buckets & Boots: Gardening Together with Children,* by Sharon Lovejoy
- *Sunflower House*, by Eve Bunting

- *The Carrot Seed,* by Ruth Kraus
- *The Curious Garden,* by Peter Brown
- *The Tiny Seed,* by Eric Carle

75. Donate a new or gently used book for adults about gardening and old issues of gardening magazines to a local library.

76. Give a container garden starter kit to a friend or family member who lives in an apartment.

First find out if container gardening is a hobby your friend or family member would enjoy. A starter kit can include: one or two colorful pots, potting soil, and a gift certificate to a local nursery for seeds or plants. Garage and moving sales are places your family can find inexpensive pots of different sizes and colors.

Include the following website address in the kit:

www.containergardeningtips.com

77. Plant several rows in your garden for a local food bank.

At **www.feedamerica.org** your family can search for food banks by zip code. Call and ask about donation guidelines.

78. Share produce from your garden with a neighbor.

79. Give a friend who is an organic gardener a packet of organic or heirloom seeds inside a handmade "thinking of you" card.

Online sources of organic and heirloom seeds include:

- The Cook's Garden: **www.cooksgarden.com**
- The Natural Gardening Company: **www.naturalgardening.com**
- Heirloom Seeds: **www.heirloomseeds.com**
- Peaceful Valley Farm Supply: **www.groworganic.com**
- Planet Natural: **www.planetnatural.com**
- Seeds of Change: **www.seedsofchange.com**
- Seed Savers Exchange: **www.seedsavers.org**

Chapter Seven: Food

"Every choice we make about food matters."

—Alice Waters

80. Take the pledge to buy local foods.

Whenever we buy directly from area farms and farmers' markets or from grocery stores that sell local products we benefit from fresh foods that haven't traveled hundreds or even thousands of miles. This simple buying practice helps preserve small, sustainable family farms that are a vital part of communities everywhere. It also boosts our local economy.

The Maine Organic Farmers and Gardeners Association, for example, estimates that if Maine residents spent just $10 a week on local foods, area farmers and the state economy would benefit by **$100 million dollars** over the course of each growing season.

Food Routes.org, which provides resources for buying local foods, encourages everyone to take the Buy Local Foods Pledge:

www.foodroutes.org/buy-local-challenge.jsp

Books About Sustainable Farming

Ecoagriculture: Strategies to Feed the World and Save Wild Biodiversity, by Jeffrey A. McNeely and Sara J. Scherr

Food for All: The Need for a New Agriculture, by John Madeley

The Farm as Natural Habitat: Reconnecting Food Systems With Ecosystems, by Nina L. Bradley

According to Sustainable Table.Org:

"Sustainable agriculture is a way of growing and raising food that is healthy to eat, doesn't harm the environment, respects workers, is humane to animals, provides a fair wage to the farmer, and supports and enhances farming communities."

81. Adopt a local sustainable farm or a farmers' market

Sustainable Table's *Eat Well Guide* is a free online directory that lists local family farms, farmers' markets, CSAs (Community Supported Agriculture farms) restaurants and other sources of sustainably grown foods.

At **www.eatwellguide.org** you can search by product, growing method, or region. Your family can also add new listings to your area's "Local Food Guide."

Local Harvest.Org also provides a nationwide map of farmers' markets, family farms, farm stands, CSAs and other sources of locally grown food: **www.localharvest.org**

Tips for shopping at a Farmers' Market:

- Vendors appreciate small bills. Take along $1 & $5 bills (quarters also help).
- Take your own bags and a cooler if you plan to buy eggs, cheese, or meat.
- For the best selection, arrive early.
- Talk with the vendors to find out which products will be available next. They may also take orders and share recipes.

If you are interested in starting a farmers' market, or you know someone who is, information is available at the USDA's Agricultural Marketing Service:

www.ams.usda.gov/FarmersMarkets

Click on "Find Farmers Market Funding Opportunities" and "Start a Farmers Market."

82. Give a family member or friend who buys or sells at a farmers' market a book on the topic.

Three gift ideas:

- *The Farmers' Market Book: Growing Food, Cultivating Community,* by Jennifer Meta Robinson and J.A. Hartenfeld
- *Local Flavors: Cooking and Eating from America's Farmers' Markets*, by Deborah Madison
- *The Farm to Table Cookbook: The Art of Eating Locally,* by Ivy Manning and Gregor Torrence

83. Sign up for a crop share from a Community Supported Agriculture (CSA) farm.

Get your name in as early as possible as CSA shares are limited in number. To find CSA farms in your area, log onto:

www.localharvest.org

84. Donate a crop share from a CSA farm to a local food bank.

An increasing number of food banks nationwide are including fresh and locally grown foods in their nutrition programs. Here in Omaha a fresh fruits and vegetables truck, like the traditional ice cream truck, drives through low-income neighborhoods providing fresh foods free of charge.

Contact a local food bank to find out whether they accept donations of CSA crop shares or produce from your backyard garden. To find CSA farms in your area, visit:

www.localharvest.org

This website maintains a database of local farms, farmers markets, CSA farms, restaurants, grocers and food co-ops searchable by zip code.

To find local food banks, visit: **www.feedamerica.org**

85. Have a farm-to-table potluck dinner party.

Serve potluck dishes made from local foods. Invite a farm family, or arrange a visit to a local farm sometime. Play old-fashioned outdoor games and sing folksongs and modern tunes about gardens and farms and saving the Earth.

Play board games, for example: Earth-Oply, Wild Seed Game, Harvest Time, Farm-Oply and Garden-Oply. Make your own Environmental Jeopardy game, with everyone contributing questions, or download Earth Day Network's free version at: **www.earthday.net/environmentaljep**

Check out, rent, or buy a DVD appropriate for all ages. Green Planet Films.Org, a nonprofit organization that promotes environmental education through film, is one of the best sources we've found: **www.greenplanetfilms.org**

For themes, search "eco dinners" or "green dinners" on the Internet and then a specific topic for links to discussion questions or guides.

Additional sources of nature and environmental DVDs:

www.amazon.com
www.bullfrogfilms.com
www.sierraclub.org

86. Donate a new or gently used food documentary to your neighborhood library, for example:

A Cow at My Table
Food, Inc.
Fridays at the Farm
One Man, One Cow, One Planet
The Future of Food
The Real Dirt on Farmer John

87. Have "Meatless Mondays."

Meatless Monday.Com's mission is to help people reduce their meat consumption in order to improve their health and that of the planet: **www.meatlessmonday.com**

According to a 2006 report from the United Nations Food and Agriculture Organization, the livestock industry generates more greenhouse gas emissions than all the cars, trucks, trains and planes in the entire world.

88. Read and discuss *Six Arguments for a Greener Diet*.

Download the free book, *Six Arguments for a Greener Diet: How a More Plant-Based Diet Could Save Your Health and the Environment,* by Michael F. Jacobson and the Staff of the Center for Science in the Public Interest. Share the link with others.

www.cspinet.org/EatingGreen/download.html

After reading the book, using the section called "Changing Your Own Diet" as a guide, begin making changes in your nutrition plan that will benefit your family's health and that of the planet.

89. Take the PB&J Pledge.

With every pb&j sandwich, or any plant-based meal your family eats instead of one based on animal products, you: fight global warming by reducing greenhouse gas emissions, conserve water, prevent water pollution, and save land.

To take the PB&J Pledge, calculate how much of an impact each plant-based meal will have, and find recipes, visit:

www.pbjcampaign.org

90. Collect vegetarian recipes.

Give everyone in your family the opportunity to choose their favorite recipes and, if they're old enough, to help prepare them. Three websites that we like are:

The Vegetarian Resource Group: **www.vrg.org**

Vegetarians in Paradise: **www.vegparadise.com**

Vegetarians Recipes.Org: **www.vegetariansrecipes.org**

Note: If your family eats meat, the Union of Concerned Scientists and other environmental organizations recommend buying it from local farmers who practice sustainable methods that are less destructive to the environment. If you buy from a grocery store, choose meats with the USDA Organic label. These come from farms that do not use synthetic fertilizers and pesticides, growth hormones, antibiotics or genetically engineered livestock seed or feed.

Veggie Teens: A Cookbook and Guide for Teenage Vegetarians, by Elyse May, Michelle May M.D., and Chef Owen May.

For information visit:

www.veggieteenscookbook.com/

91. Buy organic Fair Trade coffee.

Fair Trade certification represents sustainable farming practices, environmental stewardship, and fairness in business transactions. TransFair USA certifies coffee and other Fair Trade products imported into our nation. To locate nearby coffee shops, grocery stores and other retail businesses that sell Fair Trade coffee in your community, visit:

www.transfairusa.org

92. Choose shade grown coffee.

If you drink coffee, one of the easiest things you can do to save songbirds and other wildlife is to buy shade grown coffee from sustainable coffee farms. To learn more about how this simple buying action preserves songbird and other wildlife habitat, visit the Songbird Foundation website:

http://www.songbird.org/index_main.htm

There are shade-grown varieties of Fair Trade coffee. Ask your Fair Trade retailer about them—or wholesaler if you plan to sell coffee as a fundraiser.

93. Donate a new or gently used DVD about Fair Trade to your neighborhood public library.

This is one way your family or an environmental/social justice group you belong to can celebrate Fair Trade month in October—or any month of the year for that matter. Two documentaries we recommend are: *Buyer Be Fair* and *Black Gold.*

TransFair USA provides information about these and other Fair Trade videos, including where to order them:

www.transfairusa.org

If your family or a group you belong to registers to host a Fair Trade House Party with TransFair USA, you'll receive a free 30-minute version of *Buyer Be Fair* plus educational brochures, fact sheets, action guides, and other materials. Visit the above website address for more information.

94. Ask your congregational leaders to show *Buyer Be Fair* or *Black Gold*.

At the screening provide a list of local and national retail stores that sell Fair Trade products. Invite one or more local Fair Trade retailers to the event. Discussion guides for both videos can be downloaded at:

www.transfairusa.org

95. Donate fruit trees to developing countries.

Tens of millions of fruit trees have been planted by people in developing countries around the world through donations to Trees for Life International. This simple green gift helps recipients and their local communities become more economically self-sufficient. A single apple tree can live forty or more years and produce over 10,000 pounds of fruit: **www.treesforlife.org**

96. Parents: Give your kids the Trees for Life Adventure Gift Kit.

The $10 kit contains tree seeds, a planting carton, growing instructions and other fun stuff. With each kit Trees for Life.Org will plant 10 fruit trees in a developing country in your child's name. Click on "Learn" at:

www.treesforlife.org

97. Give the Trees for Life Adventure Kit to a classroom.

The *Trees for Life Adventure* is an educational kit from which kids in grades K-6th can learn about the role trees play in our local ecosystem and how they help solve the problem of world hunger. The teacher's guide includes hands-on activities: art and collage, placemats, leaf rubbing, role playing, poetry writing, the Fair Share Game, and identifying products made from trees.

For $1 each, students receive: a planting carton, tree seeds, a color-your-own postcard and a "Trees for Life" button. The teacher's guide is $10. Other groups that would likely appreciate this gift are a scout troop, an eco club, or a faith-based youth group working on the problem of world hunger: **www.treesforlife.org/give/catalog**

98. Provide seeds to people who can't afford to buy them.

The World Seed Fund is sponsored by the Organic Seed Alliance, a network of organic seed growers. The World Seed Fund distributes open-pollinated seeds to individuals, groups and organizations that cannot afford to buy them, including: schools, prison gardens, community and urban gardens, and agricultural projects in foreign countries.

Recipients are given instructions on how to save, share and exchange seeds in their local communities. To make a donation or to request seeds, visit: **www.seedalliance.org**

99. Provide a micro agricultural loan to an individual, family, or group in a developing country.

Micro loans enable people in impoverished communities around the world to begin enterprises that will support themselves and their families and benefit their local communities. With an agricultural micro loan an individual or a family might, for example: open a produce stand, buy livestock, raise honeybees, or plant fruit or nut trees. The minimum loan through the micro finance organization Kiva is $25.

Here's how it works: You choose the individual or group you want to lend to. When the loan is paid back, usually within 6 to 12 months, that money can go into another micro loan of your choice or you can withdraw it. Using your initial seed money, this green philanthropy project can go on for years with no additional funds required.

We have set up a lending team on Kiva.Org for agricultural and environmental-related micro loans. To find us, simply go to **www.kiva.org**, click on "community," and then in the "Search for a lending team" box enter: "Green Partnerships."

100. Give a youth, classroom, or neighborhood library a book about micro lending.

For kids:

One Hen: How One Small Loan Made a Big Difference, by Katie Smith Milway

Ricksaw Girl, by Mitali Perkins

For older youth and adults:

A Billion Bootstraps, by Philip Smith and Eric Thurman

Banker to the Poor, Micro-Lending and the Battle Against World Poverty, by Muhammad Yunus

Small Loans, Big Dreams: How Nobel Prize Winner Muhammad Yunus and Microfinance Are Changing the World, by Alex Counts

101. Youth: Read about and discuss micro lending with your family and friends.

One Hen.Org is a great source of information on micro-finance 101 for youth and how it is changing the lives of other families all around the world: **www.onehen.org**

102. Give a starter flock of chicks, ducks, or geese or a "share" of another farm animal.

The gift of a starter flock of chicks for a family in a developing country through Heifer International is $25. A flock of ducks or geese are each $20; honeybees are $30.

Another option is for your family to buy a "share" of a goat, water buffalo, heifer, sheep, llama, rabbits, or pig for $10 to $50, depending on the animal you select.

"Passing on the gift," giving someone else in the community the offspring from an animal, for instance, is a part of Heifer International's philosophy. When individuals or families pass on gifts to others and they, in turn, do the same, the entire community benefits.

Heifer International currently has over 850 active projects in 28 U.S. states and 53 countries or provinces around the world. To learn more, click on "Our Work" and scroll down to "Our Projects" at: **www.heifer.org**

103. Give a young child, a classroom, or a local library a book about Heifer International.

These two books discuss the ways in which a small gift through Heifer International changes people's lives:

Beatrice's Goat, by Page McBrier
Give a Goat, by Jan West Schrock

Chapter Eight: Water

> "Our choices at all levels—individual, community, corporate and government—affect nature. And they affect us."
>
> —David Suzuki

104. Donate a water kit to families in a developing country who are without clean drinking water.

Each kit provides enough water purification supplies for 10 families. According to the United Nations, over 1 billion people in the world do not have access to clean drinking water. Almost 6000 children die every day—more than 2 million a year—as a result of contaminated water and poor sanitation. This gift is sponsored by UNICEF.

For more information, visit **www.unicefusa.org** and click on "Donate" and then "Inspired Gifts." Your family can create your own fundraising page at this same website.

105. Choose seafood that helps save ocean environments.

One simple thing your family can do to protect oceans is to choose seafood that has been caught or farmed in ways that do not threaten species with extinction or severely damage their habitat. On the website, Seafood Watch.Org you can print free pocket guides to carry with you to restaurants and grocery stores. The guides list best seafood choices, good alternatives and those to avoid: **www.seafoodwatch.org**

106. Collect change in your Green Philanthropy Bank for Ocean Conservancy.

"Going green starts with living blue."

—Ocean Conservancy.Org

Ocean Conservancy is the largest and oldest organization in the United States dedicated to protecting ocean environments. Besides collecting and donating funds from your Green Philanthropy Bank, here are four more things your family can do to help conserve oceans:

- Sign up for the *Ocean Conservancy eNewsletter.*
- Join Ocean Conservancy.Org. The *Ocean Conservancy Magazine* is free for a membership at the $25 level or higher.
- Take part in the Ocean Action Network and tell others about it.
- Volunteer for Ocean Conservancy's International Coastal Cleanup of rivers, lakes and beaches.
  Select "Take Action" from the menu bar and under "International Coastal Cleanup" click "Think globally, act locally:" **www.oceanconservancy.org**

107. Read and discuss 50 *Ways to Save the Ocean*, by David Helvarg.

Choose one or more additional green philanthropy actions or projects than those previously mentioned to help protect ocean environments. When you finish the book, pass it on to another family or donate it to your neighborhood library.

Resources for youth ages 8-12: Project WET publishes two inexpensive activity booklets: *Discover Oceans* and *Discover Coral Reefs*. $1.25 each. For more information, at **www.projectwet.org** click on "Store" and then "Children's Activity Booklets."

108. Give a teacher Cousteau in the Classroom.

The *Cousteau in the Classroom* water education program provides enough materials for 30 students, including copies of the bi-monthly award-winning magazine *Cousteau Kids*. Teaching guides with student activities for each issue are available online at: **www.cousteaukids.org**

The mission of the Cousteau Society is to educate people ". . . to understand, to love and to protect the water systems of the planet, marine and fresh water, for the well-being of future generations."

109. Parents: Give your kids a membership to Cousteau Kids.

Membership includes a subscription to the award-winning magazine, *Cousteau Kids*, for ages 8-12, and online access to related educational information and activities. To learn more, log onto **www.cousteau.org** and click on "Cousteau Kids."

110. Give a pack of Project WET water education booklets to a classroom, kids eco club, scout troop, or your own family.

Project WET publishes a series of colorful, 16-page activity books for ages 8 to 12 that include educational stories, games, and science experiments about water. You can order them online at **www.projectwet.org** by clicking on "Store" and then "Children's Activity Booklets." Or you can call 1-866-337-5486.

Booklet titles include: *Celebrate Wetlands, Conserve Water, Discover Ground Water and Springs, Discover Storm Water,* and *Watershed Protection.*

111. Parents: Practice green housekeeping and teach it to your kids.

This is one of the easiest things you can do to protect our waterways. We recommend this how-to book by Annie Berthold-Bond: *Clean and Green: The Complete Guide to Nontoxic and Environmentally Safe Housekeeping.* It has 485 recipes that are healthier for your family and the Earth.

Alternatively, your family can create your own recipe book or booklet in PDF format. Search online for "green cleaning recipes." Collect and save the ones from your favorite environmental organizations.

This is a fun and educational project for eco clubs, school

sustainability programs, family nature clubs and other groups. Putting it in PDF format saves paper and makes it easy to send.

112. Think outside the bottle.

That's the name of a major campaign sponsored by Corporate Accountability International, calling upon people worldwide to give up bottled water for the common good. Some companies take communities' water and privatize it despite public opposition all around the world.

There is a growing counter movement against privatization. Your family can help keep water public and not a for-profit commodity simply by refusing to buy or drink bottled water. Thinking outside the bottle means asking whose community the water comes from and who benefits from its sale: **www.thinkoutsidethebottle.org**

Check out *The Story of Bottled Water* at Annie Leonard's website: **www.storyofstuff.com**

113. Take back the tap.

That's the name of another major campaign sponsored by Food and Water Watch, encouraging people all over the world to take the pledge to give up bottled water and ask restaurants not to serve it either:

www.takebackthetap.org

Food and Water Watch provides a guide to water filtration at:

www.foodandwaterwatch.org/water/pubs/water-guides/filtration-guide/water-filtration-guide

<u>More information about bottled water:</u>

Billions of plastic water bottles are thrown into landfills every year: **www.container-recycling.org**

Pepsi's Aquafina and Coca-Cola's Dasani come from municipal water supplies. – Reuters

The U.S. Environmental Protection Agency's standards for tap water are more stringent than the Food and Drug Administration's standards for bottled water.

–*Is Bottled Water Better?* Union of Concerned Scientists

114. Parents & Older Youth: Watch and discuss a DVD about "water as a human right."

Donate it to your congregational or local library afterwards. We recommend these three documentaries:

Blue Gold
FLOW (For the Love of Water)
The Water Front

You can order a copy of *The Water Front* and download a screening guide at:

www.waterfrontmovie.com

115. Sign the petition calling on the United Nations to add Article 31 to the Universal Declaration of Human Rights.

Article 31, the "water as a human right" article, states that:

"Everyone has the right to clean and accessible water, adequate for the health and well-being of the individual and family, and no one shall be deprived of such access or quality of water due to individual economic circumstance."

To sign the petition log onto: **www.article31.org**

Resource for Youth and Parents: Read and discuss the thirty articles in the Universal Declaration of Human Rights with your family. Food and shelter are covered under Article 25. The following website provides information specifically geared toward youth:

www.un.org/events/humanrights/udhr60/ youth.shtml

116. Collect change in your Green Philanthropy Bank for Ryan's Well Foundation.

In 1998 after Ryan Hreljac learned from his first-grade teacher that people were dying from a lack of clean water, he earned and saved $70 to help build a well. What began as one youth's single act of green philanthropy has since grown into a foundation that has raised millions of dollars and brought over 500 water and sanitation projects to more than 600,000 people in 16 countries.

www.ryanswell.ca/

<u>**You can read about Ryan in a book co-authored with other young green philanthropists called: *What's It Like Living Green? Kids Teaching Kids, by the Way They Live.***</u>

Chapter Nine: Urban Habitats, Wild Places, and Wildlife

> "There is hope if people will begin to awaken that spiritual part of themselves, that heartfelt knowledge that we are caretakers of this planet."
>
> —Brooke Medicine Eagle

117. Parents and older youth: Become certified Master Naturalists.

Almost thirty states now have a Master Naturalist program, similar to the Master Gardener program. It's an opportunity for volunteers to learn more about local ecosystems and biodiversity and to work with others who are passionate about conserving our natural resources.

Participants receive classroom instruction and do field-work alongside scientists, professional land managers, university faculty and other environmental experts. Depending on a state's volunteer needs: Certified Master Naturalists can lead public education programs, engage in conservation work in local natural areas, staff nature centers and conduct citizen science.

To find out if your state has a program, search online for:
"(Your State) Master Naturalist Program."

118. Youth: Do chores to earn money for your family to purchase a year's pass to the National Parks and Federal Recreational Lands.

This makes a great gift for families that frequently visit parks and federal lands. For more information, log onto:

www.nps.gov/fees_passes.htm

119. Sponsor a wild place.

When your family sponsors a wild place through the Sierra Club, you help preserve it and the wildlife that inhabits it. Shortly before this book went to print, the following sponsorships, starting at the $20 level were available:

Glacier National Park
Yosemite National Park
Grand Canyon National Park
Giant Sequoia National Monument
Arctic National Wildlife Refuge
Yellowstone National Park
Everglades National Park
Acadia National Park

To find out more about this conservation program, including whether any new wild places have been added to it, log onto: **www.sierraclub.org/SponsorWildPlaces**

<u>Learn about Earth's habitats and the plants and animals that live in each of them at this website:</u>

www.nationalgeographic.com/environment/habitats/

120. Adopt an acre.

The Nature Conservancy began the Adopt an Acre program in 1991. Since then families, individuals, school children, and other green philanthropists have protected and conserved over 600,000 acres of rainforests all around the world. With an adoption of $50 an acre, your family can choose to conserve forests in the United States, lush rainforests in other parts of the world, or one of the coral reefs:

www.nature.org/adopt/

121. Help save BioGems.

BioGems are ". . . irreplaceable remnants of wilderness that provide sanctuary for endangered wildlife, curb global warming, preserve biodiversity and enrich the lives of local people." Save BioGems is a campaign sponsored by the Natural Resources Defense Council.

One of the easiest things your family can do to is to send already-prepared email messages calling on decision makers in government and business to save BioGems for future generations. More than half a million people are BioGems Defenders. Since the network was first launched in 2001, over ten million messages have been sent. To learn more about the program, log onto:

www.savebiogems.org/about

122. Collect funds in your Green Philanthropy Bank to help save BioGems.

For more information, visit **www.savebiogems.org**

123. Parents: Teach your kids how to write a letter or an email to decision makers.

These wikiHows below are helpful resources. Go to **www.wikihow.com** and in the search box enter: "How to Write a Well-Organized Letter" or "How to Address a Letter to a Government Official"

In addition to sending letters or emails advocating for different issues, also teach your kids to write and thank government and business leaders for their work.

124. Download and save the free interactive book: *Thoreau's Legacy: American Stories About Global Warming*

This project of the Union of Concerned Scientists is a collection of writing and art from 67 individuals about "beloved places, animals, plants, people, and activities at risk from a changing climate." Your family will learn what each person is doing to conserve what they love and some actions you can take as well. To download a free copy, log onto: **www.ucsusa.org/americanstories**

125. Adopt a wildlife animal

> "How do we care for all the children of all the species for all time?"
>
> —William McDonough

A number of local and national wildlife conservation organizations sponsor symbolic adoptions. These include the National Wildlife Federation, World Wildlife Fund, and Defenders of Wildlife. These adoptions are a way people of all ages can learn about and help save threatened and endangered animals, one gift at a time. Many adoptions are available at the $20 to $30 level.

To learn more about each adoption program, visit these websites:

Defenders of Wildlife: **www.wildlifeadoption.org**

National Wildlife Federation

www.nwf.org/adoptioncenter

World Wildlife Fund: **www.worldwildlife.org**

126. Give a young child the book, *Will We Miss Them?*

The book is about endangered wildlife animals and was written by Alexandra Wright when she was eleven years old. It makes a great companion gift to one of the wildlife adoptions previously described.

Two more books about endangered wildlife:

Can We Save Them?: Endangered Species of North America, by David Dobson

Where Have All the Pandas Gone?, by Melvin and Gilda Berger

Wildlife Guides:

The website **www.enature.com** has field guides to 5500 wildlife species, including information on threatened and endangered species. The "Ask an Expert" feature allows you to email a question about wildlife.

Chapter Ten: Saving Forests

"If we do nothing else, save the rainforest."

—Al Gore

127. Youth: Learn about rainforests and how you can help save them.

The Rainforest Alliance has many projects and activities on its website, including: rain forest stories, games, coloring book, animal facts, and ways young people everywhere can help save rainforests:

www.rainforest-alliance.org/education

Another great website with educational resources on rainforests, including links to *Journey Into Amazonia* and *Passport to the Rainforest:*

www.rainforesteducation.com

128. Plant a rainforest tree for $10.

The Revive a Rainforest campaign is sponsored by the Natural Resources Defense Council. Each $10 your family donates will plant a tree in an imperiled rainforest.

For more information, click on "Revive a Rainforest" at:

www.savebiogems.org/giftsfromthewild/index.html

129. Collect funds in your Green Philanthropy Bank to save rainforests.

The Protect-an-Acre Program, sponsored by the Rainforest Action Network, provides small amounts of money, or grants, to indigenous people living in rainforests. These grants enable them to obtain title from the government for the land they inhabit.

Ownership gives indigenous communities control over their land. They can remain on it for as long as they want, and their trees, gold, oil and other natural resources cannot be taken or harmed.

The Protect-an-Acre Program is funded by families, individuals, classrooms, schools and other groups. For more information, visit:

www.ran.org/new/kidscorner/

Download the *Kids Action Toolkit* to discover more ways your family can help save rainforests:

www.ran.org/new/kidscorner/

130. Donate a book about rainforests to a local school or public library.

World Rainforest Week, sponsored by the Rainforest Action Network, is in October each year. Any day of the year, however, is a good time to celebrate rainforests and their rich biodiversity. For more information about World Rainforest Week and the Rainforest Action Network, visit:

www.ran.org

Here are 7 gift ideas:

- *A Rainforest Habitat,* by Molly Aloian and Bobbie Kalman
- *A Walk in the Rainforest,* by Kristin Joy Pratt-Serafini
- *The Great Kapok Tree: A Tale of the Amazon Rain Forest,* by Lynne Cherry
- *The Rainforest Grew All Around,* by Susan K. Mitchell
- *The Remarkable Rainforest: An Active-Learning Book For Kids,* by Toni Albert
- *Tropical Rainforest,* by Donald Silver
- *Welcome to the Greenhouse,* by Jane Yolen

131. **Offset your reading footprint by planting 1 tree for $1 for every book you buy that's printed on non-recycled paper.**

The nonprofit organization Eco-Libris is working toward the goal of persuading all publishers to print books on recycled paper. Until that day arrives, they provide a way for readers to support sustainable forests in developing countries and to promote green publishing.

For any number of books you purchase that are printed on non-recycled paper, you can choose to have 1 tree planted for $1 by an Eco-Libris planting partner: International

Rainforest Alliance, Ripple, or Sustainable Harvest. For more information, their website address is:

www.ecolibris.net

132. Plant more $1 trees through these two conservation organizations:

American Forests:

www.americanforests.org

Nature Conservancy's "Plant a Billion Trees:"

www.plantabillion.org

133. Support the Green Press Initiative.

Here are three things your family can do to support the work of the Green Press Initiative:

- Become informed. Every year the U.S. newspaper industry uses 95 million trees, the book industry 30 million. Many are from old growth and endangered forests. The paper industry emits almost 750 million tons of CO_2 annually, or the equivalent of the yearly emissions of more than 136 million cars.
- Collect funds in your Green Philanthropy Bank for the Green Press Initiative.
- Write to your local newspaper and favorite book publisher and let them know that your family supports environmentally friendly publishing. First, log onto **www.greenpressinitiative.org** and check

to see if they're listed among the Green Press Initiative's "Industry Leaders."

Thank them if they already have an existing policy of sustainable printing practices. If they don't, encourage them to make the transition and let the Green Press Initiative know about it. For sample letters contact: **kelly@greenpressinitiative.org**

To learn more about what your family can do to support green publishing, go to **www.greenpressinitiative.org** and click on "Action Steps."

134. Help conserve our nation's ancient redwood forests.

Since 1918 the Save the Redwoods League has conserved over 180,000 acres of ancient redwood forests and developed 59 redwood parks and reserves for everyone to enjoy. One of the first steps your family can take to conserve the forests is to learn more about the League's work and about the ancient redwood trees. A reading list of books for youth is posted on their website. These include:

- *A Tree Called Moses*, by Laura Nelson Baker. Ages 4 and up.
- *While a Tree Was Growing,* by Jane Bosveld. Ages 4-8.
- *The Ever-Living Tree,* by Linda Vierira. Ages 4-8.
- *Redwood National Park: Forest of Giants,* by Neil Purslow. Ages 8-12.
- *Ancient Forests: Discovering Nature,* by Margaret Anderson, et. al. Ages 8-13.
- *Julia Butterfly Hill: Saving the Ancient Redwoods,* by Rachel Lynette. Ages 9-12.

- *Legacy of Luna,* by Julia Butterfly Hill. Young Adult.
- *The Wild Trees: A Story of Passion and Daring,* by Richard Preston. Young Adult.

For the complete list, visit:

www.savetheredwoods.org/kidsbooks

135. Download free books and other publications online.

Your family can download 30,000 free electronic books (ebooks) through Project Gutenberg and 100,000 free ebooks through their partners, affiliates, and resources. You can download ebooks to your PC, iPhone, Kindle, Sony Reader or other portable device. To check out Project Gutenberg's catalog and links to other free ebooks, visit:

www.gutenberg.org

136. Collect change in your Green Philanthropy Bank to help Project Gutenberg provide green books to people worldwide.

The mission of the Project Gutenberg Foundation is to provide downloadable books and other literary works to people all over the world free of charge or at the lowest cost possible. For more information, or if your family or a group you belong to would like to fundraise for Project Gutenberg, log onto: **www.gutenberg.org**

137. Parents: Sign Forest Ethics' petition to create a National Do Not Mail Registry.

Similar to the National Do Not Call Registry, which has given families freedom from unwanted calls, a National Do Not Mail Registry would allow us to "take back our mail boxes:" **www.donotmail.org**

It takes millions of trees every year to produce our nation's enormous sea of junk mail. You can use Forest Ethics' online tool to get rid of your own:

www.donotmail.org/form.php?id=50

138. Give 5 friends the above links to Forest Ethics' online petition and tool to get rid of junk mail.

Chapter Eleven: Additional Natural Resources Conservation

"A non-renewable resource is a natural resource which cannot be produced, re-grown, regenerated, or reused on a scale which can sustain its consumption rate."

—from *Wikipedia*

139. Contribute approximately $14 each month from your Green Philanthropy Bank to purchase good quality carbon offsets.

The three steps for your household to become "carbon neutral" are: reduce your energy use as much as possible, offset the rest, and then repeat.

We recommend buying carbon offsets from providers that: 1. fund quality carbon-busting projects and 2. have been certified by the International Carbon Reduction and Offset Alliance **www.icroa.org** or Green-e Climate **www.green-e.org/climate**

Native Energy is an example of a certified provider. It funds projects that benefit schools, family farms, and Native American communities. The types of projects developed through the purchase of carbon offsets by individuals and businesses from Native Energy are:

renewable energy, farm methane power, solar, biomass, and transportation efficiency. The current featured project is the Greensburg Wind Farm.

Native Energy estimates that the average U.S. household emits 12 tons of carbon dioxide each year. Carbon offsets can be purchased from Native Energy for $14 per ton, which offsets the average home's power and heat for 1 month. For more information, go to:

www.nativeenergy.com

For lists of other certified carbon offset providers and their carbon-busting projects, visit the websites of the two certifiers mentioned above.

To learn more about carbon offsets in general, log onto the website of Carbon Offset Research & Education:

www.co2offsetresearch.org/consumer

140. Youth: Do a science fair project on natural resources conservation in which your family can become involved.

Check out this website for science fair project ideas and information, including links to other sites:

www.all-science-fairprojects.com

The EPA's Teaching Center has numerous educational resources on conservation for students and teachers, many of which are downloadable: **www.epa.gov/teachers**

Click on "Conservation"

141. Give a bag of green gifts to someone who can't afford to "go green."

This makes a great gift for members of your extended family or friends who live on a fixed or low income. It helps them save money and benefits the environment at the same time. Earth Day, Friendship Day, Valentine's, or birthdays are all good times to give this gift. Look around their houses to see which items they might be able to use.

A poll taken by researchers at Yale and George Mason Universities shows that people want to make their homes more energy efficient but fear the cost.

Things to include in a green-gifts bag:

- large containers for storing bulk purchases
- permanent coffee filter
- green cleaning supplies
- recipes for money-saving green cleaners
- programmable thermostat
- battery-charger and a set of rechargeable batteries
- energy saving light bulbs
- solar lights
- motion sensor lights
- reusable bags
- LED nightlight
- faucet aerators
- tire gauge
- clothes line or drying rack

142. Collect funds in your Green Philanthropy Bank to help provide solar power to people who live without electricity.

The Solar Electric Light Fund is a nonprofit organization whose mission is to provide solar power to as many as possible of the nearly 2 billion people in the world who live without electricity. It partners with many organizations including: Partners in Health, Millennium Villages Program, Make It Right Project, Jane Goodall Institute, Institute for Sustainable Energy and Development, United Nations, Habitat for Humanity International, and the U.S. Department of Energy.

It has brought solar electrification projects to more than 15 countries, including the United States. The Navajo Nation and the city of New Orleans are among the many beneficiaries and partners of the program: **www.self.org**

143. Give a Solar Cooker Kit to a family housed in a refugee camp.

Simple low-tech solar cookers are protecting the lives of women and girls housed in refugee camps in war-torn countries. Without the cookers they would have to go outside the camps where there is no protection and gather firewood.

They are taught how to use the solar cookers and how to make them, which generates income for their families. These tools are cleaner, eliminating wood smoke from the surrounding air. Just two of them save one ton of trees every year.

A family is given two solar kits: one for cooking maize and the other for water, sauce, or vegetables. Water can be boiled on the solar cookers, which kills disease-causing water-borne microbes.

The Solar Cooker Project was created by the nonprofit organization Jewish World Watch. Over 500 organizations, churches and synagogues, as well as individual green philanthropists, have donated funds for the purchase of solar cookers. Because of their health and ecological benefits their use is extending to an increasing number of people worldwide.

www.solarcookerproject.org

144. Youth: Learn how you can help your parents save energy and curb global warming at the same time.

On the U.S. Department of Energy website for kids you'll discover numerous ways to save energy, learn about the different kinds of renewable energy, play games, and take a test to find out what you know about energy:

www.eere.energy.gov/kids/

Make a list of several things you can do to help your parents save energy and money and curb global warming. Post it on your bulletin board or refrigerator.

A Free Resource for Parents

The *Consumer's Guide to Energy Efficiency and Renewable Energy* shows you how you can save energy in your home, vehicle, and workplace:

www.eere.energy.gov/consumer

145. Give the leaders in your religious organization the link to the *Energy Star Guide for Congregations.*

The guide is a free downloadable resource that will enable your congregation to save energy and money and curb global warming. Enter the title into your favorite search engine: *Putting Energy Into Stewardship: Energy Star Guide for Congregations.*

146. Parents and Older Youth: Watch and discuss the documentary *Kilowatt Ours.*

As you watch it, write down the actions your family can take in your home, business or community and add them to your green philanthropy plan.

This award-winning documentary explores our use of nonrenewable sources of electricity and their adverse health and environmental impacts, including mountain-top removal, global warming, and asthma. It also discusses the benefits of renewable and sustainable energy and actions we can take in our homes, businesses and communities to build a "net zero nation."

For more information about *Kilowatt Ours* or to view Jeff Barrie's 3-minute video called, "I Dream of a Green Home," visit: **www.kilowattours.org**

147. Give a booklet of bus tickets or tokens to someone who needs them.

Lack of affordable transportation is a common barrier for young or low-income people who seek education and employment. If you know someone in your congregation, neighborhood, or your own family who regularly rides the bus to work or school, bus tickets or tokens make a thoughtful and useful gift.

Chapter Twelve: Green Philanthropy Fundraisers

These fundraisers are, themselves, green philanthropy projects. Some your family can do alone, while others can involve members of a local environmental group to which you belong.

148. Give it up for green philanthropy.

Discuss with your family that half the people in the world live on less than two dollars a day. Brainstorm on the things your family can buy for that amount of money.

Make a list of what you each would be willing to give up for a week or however long in order to save money for a green philanthropy gift or project. For example: take your lunch to work or school instead of buying it; watch a movie at home and make your own popcorn; check out a book from the library instead of buying it; prepare a vegetarian meal instead of one based on meat, etc.

Groups can also use this activity as a way to raise funds for green philanthropy.

149. Collect printer cartridges and cell phones to benefit a local environmental organization.

Your family can help raise money for a local environmental group by collecting laser printer cartridges and cell phones and recycling them with the Green School Project. Schools and other nonprofit organizations qualify for this fundraising program.

Discuss the idea with the environmental group your family wants to support and how you might work together on this community service project. To learn more about the Green School Project, log onto:

www.greenschoolproject.com

150. Clean up a park or part of a trail as a group fundraiser.

Local chapters of Keep America Beautiful will donate money to nonprofit groups for picking up litter from a park or trail, and many of them will even provide bags, gloves, and tools.

The Keep America Beautiful website has a map locator of affiliates: **www.kab.org**, or you can look up "Keep (the name of your city) Beautiful" in your telephone book.

151. Build and sell good nest boxes.

This is a fun way for youth groups and their families to raise money for a green philanthropy project or gift. On the following website scientists at the Cornell Lab of Ornithology discuss the general features of a good nest box and provide detailed instructions on how to build nest boxes for different species of birds:

www.birds.cornell.edu/nestinginfo/nestboxref/construct

152. Build a Leopold bench as a fundraiser or a community service project.

This fundraising/community service idea is part of the Leopold Education Project. To download and print the instruction sheet for building a Leopold bench, log onto:

www.lep.org/lep_bench/

Your family or a group you belong to can build one or more benches as a community service project and donate it to a school or community garden, park, wildlife refuge, or a library among other places. Alternatively, it can be sold or raffled as a fundraiser.

153. Collect and sell a booklet of recipes from local farmers' markets.

Include information about the vendors and their products. You may want to add one or more locally grown items to your fundraiser, for example: organic popcorn, honey, garden and flower bed plants.

154. Prepare a lunch or dinner from foods grown on a local sustainable farm.

This is also a means to raise awareness about the ways buying locally grown foods benefit farmers, communities, and the environment. Invite a farm family to your event, or arrange a field trip to a local farm as a follow-up to the lunch or dinner. Involve the youth in your group and in the community:

Check Rodale's: *Farming for Credit Directory* to find out if there is a student farm in your area:

www.rodaleinstitute.org/ffc_directory

Free Downloadable Presentation Toolkits

Sustainable Table provides the following resources on its website. (Click on "Presentation Kits"):

Sustainable Food Kit
Factory Farm Kit
Create Your Own Kit

www.sustainabletable.org

Toolkit for student organizers:

Buy Local Food and Farm Toolkit: A Guide for Student Organizers, Oxfam America:

www.sustainabletable.org/schools/docs/food_farm_toolkit.pdf

155. Sell Fair Trade coffee, tea, and/or chocolates.

Congregations were among the first supporters of the Fair Trade movement in the United States. Selling Fair Trade coffee and tea remains today one of the most widely supported interfaith environmental and social justice fundraising projects in our nation.

Chocolates are another popular Fair Trade fundraiser, especially during holidays, Valentine's Day, and Mother's Day. Equal Exchange is a wholesaler for Fair Trade mini chocolates: **www.equalexchange.coop**

Create educational posters to raise awareness about Fair Trade; include information about your fundraiser. During your fundraiser show a short documentary about Fair Trade:

A thirty-minute version of *Buyer Be Fair* and educational materials are available for free for families, individuals, and groups that host a house party. For more information, visit: **www.transfairusa.org** and click on "Resources."

The short documentary *Chocolate Country* won the Grand Jury Prize for its category at the 2007 Seattle International Film Festival:

www.chocolatecountryfilm.com

At **www.transfairusa.org** your family or group can search for other wholesalers of Fair Trade fundraising products besides Equal Exchange. Also, look under "coffee" in your telephone directory for local wholesalers.

More information about Fair Trade is available at these websites:

Fair Trade Federation: **www.fairtradefederation.org**

World Fair Trade Organization: **www.wfto.com**

156. Youth: Hold a family-sponsored Free-Rice Game-athon.

This is a fun way for middle-school age or older youth to raise money for green philanthropy, provide food for people in developing countries, and increase your vocabulary all at the same time. Talk it over with your parents and if they approve, ask them and possibly your grandparents and other extended family members to sponsor you. Tell them about the Free Rice website:

www.freerice.com

Decide how long your game-athon will last and a suggested donation amount, for example: $1 for every 2000 grains. Tell your sponsors how many thousands of grains you plan to generate within your time frame: 16,000 grains will raise $8 from each family sponsor.

Green philanthropy projects and gifts that will help solve world hunger include: a micro loan through Kiva for an agricultural or environmental enterprise; providing fruit trees ($1 each) through Trees for Life to a developing country; giving a starter flock of chicks, ducks or geese through Heifer International.

157. Collect and sell gently used books to raise funds for an organization that conserves forests.

These include, among others: Rainforest Action Network, Nature Conservancy, Greenpeace, Friends of the Earth, American Forests, Forest Ethics, Sierra Club, World Wildlife Fund, and Rainforest Alliance. Research and suggest other forest-conservation organizations and then vote on one.

Sell the books at a garage sale or to a shop that deals in used books. Check the telephone directory for stores in your area.

158. Collect used cell phones to benefit a national environmental organization.

Earthworks Action.Org is just one of many nonprofit organizations that partners with Collective Good.Com to responsibly recycle used cell phones. Earthworks estimates that 500 million cell phones have either been thrown away or sit unused in our homes and that 130 million more are added to landfills every year.

Earthworks receives $1 for every cell phone you recycle through Collective Good.Com. The nonprofit organization uses the funds raised for public education and to promote recycling efforts. For more information log onto:

www.recyclemycellphone.org/recycle.cfm
and **www.collectivegood.com**

Other environmental groups that you can choose to support by recycling cell phones through Collective Good include: Environmental Learning Center, Captain Planet

Foundation, Earth Share, and Red Jellyfish, a Nature Conservancy Adopt an Acre program.

Collective Good honors the "Electronics Recyclers Pledge of True Stewardship." An essential part of that pledge involves not shipping our nation's electronic waste to Third World Countries.

159. Sell Forest Choice pencils to raise money and awareness about FSC certified products.

Earth Day, or any day of the year, is a good time to raise awareness about the Forest Stewardship Council and where consumers can buy FSC-certified school supplies, furniture, building materials and other products locally. Forest Choice pencils are one FSC-certified product that ecology clubs, schools and other groups can purchase at wholesale for a fundraiser.

Forest Choice defines a wholesaler as "any group, organization, or company that will be purchasing these FSC-certified pencils for use or resale." For more information, visit: **www.forestchoice.com/wholesalers**

The Forest Stewardship Council maintains a database of retailers that sell building materials, furniture, paper and other FSC products:

www.fscus.org/productsearch/retailers/

One way your family or group can provide this information is to print one or two sentences about the FSC database and the above website address on small strips of paper; punch a hole in one corner with a paper punch, and using yarn or string tie one to each pencil.

160. Sponsor a green raffle.

A raffle is a fun way for environmental groups to raise money and awareness about green issues. Check your state's rules for fundraising raffles to find out if you'll need a permit.

Here are five themes for fundraising raffles:

1. **Green Books & DVDs**

Gently used or new books and DVDs about nature and the environment appropriate for different age levels: pre-school, elementary, middle school, high school and adult.

2. **Natural Resources Conservation:**

Products that save resources, for example: tire gauge, energy saving light bulbs or a gift certificate to a store that sells different sizes, low flow showerhead, programmable thermostat, solar lights, 100% post consumer recycled print paper, permanent coffee filter, cloth napkins, information card on how to get rid of junk mail, etc.

3. **Green Cleaning Supplies:**

Earth-friendly cleaners in a tote. Refrigerator posters with recipes for green cleaners can be given as a free gift to everyone.

4. **Local Foods:**

Non-perishables such as honey, organic popcorn, jam and jelly; a gift certificate to a local farmer's market; a new or gently used food documentary.

Provide information for everyone on where to find farmers markets, local farms that sell directly to consumers, how buying local foods benefits people and the earth, etc.

5. **Green Philanthropist:**

For example:

$50 or so of local foods donated to a community food bank in the winner's name; 50 trees planted in American Forests' Global ReLeaf Campaign; 4 Solar Cooker Kits donated to the Solar Cooker Project.

RESOURCES

The following websites provide information about volunteer opportunities for environmental community service. Most of them also provide toolkits and other materials to support individuals, families, and groups as you develop a project of your own. Throughout our book we've also included numerous resources related to specific gifts, actions, activities and projects:

Community Garden.Org: Volunteer in a local community garden or help start one. To find a community garden in your area by the name of the garden or by zip code, city or state, visit: **www.communitygarden.org**

Earth Day Network: Register your community service project online and recruit volunteers if you need them. The toolkit, *Earth Day in a Box* tells you how to plan and sponsor a project: **www.earthday.net**

The Environmental Protection Agency provides free, downloadable publications about community service projects including: *Environmental Protection Begins With You, Linking Girls to the Land, Service-Learning Beyond the Classroom,* and *Volunteer for Change:*

www.epa.gov/teachers/community-svc-projects.htm

EPA's Adopt Your Watershed Program: Find watershed groups in your area by zip code: **www.epa.gov/adopt/**

The website maintains a database of 2600 groups nationwide.

Global Youth Service Day: The largest service event in the world that takes place every year in over 100 countries:

www.gysd.org

Groundwater.Org: Community service project ideas for youth and their families:

www.groundwater.org/ta/kidsprotect.html

Local Harvest.Org: Search for CSA farms in your area for volunteer opportunities on small sustainable farms:

www.localharvest.org

National Parks Service: There are almost 400 national parks in the United States with many volunteer service projects. Read the "Volunteers-In-Parks" brochure to learn more: **www.nps.gov/getinvolved/volunteer.htm**

National Public Lands Day: Search for volunteer opportunities in your area: **www.publiclandsday.org**

National Wildlife Federation's Schoolyard Habitats Program: Volunteer in or help start a schoolyard habitat:

www.nwf.org/schoolyard

Nature Conservancy: Contact your local chapter about volunteer opportunities: **www.nature.org/volunteer**

Ocean Conservancy's International Coastal Cleanup of rivers, lakes and beaches: Log onto:

www.oceanconservancy.org

Select "Take Action" from the menu bar. Under "International Coastal Cleanup" click "Think globally, act locally."

Serve.Gov: Search for local volunteer opportunities for environmental community service. You'll also find toolkits on this same website for creating and registering your own project: **www.serve.gov**

Sierra Club: Contact your local chapter about volunteer opportunities: **www.sierraclub.org/chapters**

Teens for Planet Earth.Org provides online interactive guides to support teens as they develop an environmental community service project:

www.teensforplanetearth.org

U.S. Fish and Wildlife Service – Wildlife Refuge Projects: There are over 530 National Wildlife Refuges in the United States. To find out if there is one near you, visit: **www.fws.gov/refuges**

Youth Service America.Org: Geared toward youth ages 5 to 25, this organization provides the tools, information and resources for planning, designing and completing a service project locally, nationally or globally: **www.ysa.org**

Global Youth Service Day: **www.gysd.org**

Free or Inexpensive Information Sources

The following websites are sources of a wide range of multimedia materials on nature and the environment that families, individuals and groups may find helpful in planning a presentation or a community service project.

The Library of Congress has put together a collection of multimedia resources on nature and the environment that the public can access for free, including: Webcasts, exhibitions, presentations, and classroom materials:

www.loc.gov/teachers/classroommaterials/ themes/nature

The EPA's Environmental Education Center has numerous free and inexpensive resource materials. Many of them can be downloaded: **www.epa.gov/teachers**

The EPA Student Center also has many free downloadable resources available on conservation, climate change, ecosystems, water, and other topics:

www.epa.gov/students

You'll find EPA resources on sustainability at:

www.epa.gov/greenkit/student.htm

The North American Association For Environmental Education provides links to organizations involved in a wide range of issues including: consumption and lifestyle, endangered species, forests and agriculture, schoolyard ecology and many others:

www.eelink.net/pages/Environmental+Information-+General

GREEN PHILANTHROPY PLAN

Write the gifts, projects, and actions that you plan to complete.

	#	Target Date	Notes
Gifts			
Projects			
Actions			

GREEN PHILANTHROPY PLAN

Write the gifts, projects, and actions that you plan to complete.

	#	Target Date	Notes
Gifts			
Projects			
Actions			

GREEN PHILANTHROPY PLAN

Write the gifts, projects, and actions that you plan to complete.

	#	Target Date	Notes
Gifts			
Projects			
Actions			

GREEN PHILANTHROPY PLAN

Write the gifts, projects, and actions that you plan to complete.

	#	**Target Date**	**Notes**
Gifts			
Projects			
Actions			

ABOUT THE AUTHORS

Helen Deffenbacher is a volunteer for the Green Omaha Coalition and a founding member of the Omaha chapter of Slow Food USA. She and her husband Ken, a retired professor, enjoy their roles as ecostewards in their local community and opportunities to work with people of all ages, including their two older grandkids Justin and Alexis with whom Helen has co-authored *Green Philanthropy For Families.* A number of their gifts, projects and actions are contained in the book.

Helen has a master's degree in education.

Justin Deffenbacher received the youth social justice service award in his church's eight-state region for co-founding an ecology club run by and for youth. When he was a member of a community service group in his school called Project Citizen, his idea of providing healthier foods in the schools was chosen by club members and his group was among several that presented their projects to Governor Dave Heineman. He enjoys volunteering for service projects, especially prairie restoration and preservation. Currently he's collecting used books for a local Reach Out and Read program.

Alexis Deffenbacher joined the youth ecology club in her church when it was first formed and signed up for various projects including selling Fair Trade chocolates, making posters to educate church members about Fair Trade, and volunteering for the club's "Adopt-a-Trail" cleanup at Standing Bear Lake. She has volunteered for the Omaha Humane Society and for the Henry Doorley Zoo's project of creating a butterfly garden on the zoo's grounds. She is presently collecting eyeglasses for Lions' Recycle for Sight program.

NOTES

NOTES

LaVergne, TN USA
24 May 2010
183737LV00005B/1/P

9 780615 331157